Quilts
from Simple Shapes

TRIANGLES

Mary Coyne Penders

Quilt Digest Press

Copyright © 1991 by Mary Coyne Penders. All rights reserved. Published in the United States of America by The Quilt Digest Press.

Editorial and production direction by Harold Nadel.
Book and cover design by Kajun Graphics.
Photography by Sharon Risedorph.
Computer graphics by Kandy Petersen.
Typographical composition by Jet Set.
Printed by Nissha Printing Company, Ltd., Kyoto, Japan.
Color separations by the printer.
Home graciously lent by Margaret Peters.
Construction specifications by Lenore Parham.

Special thanks to Cotton Patch, Lafayette, California; G Street Fabrics, Rockville, Maryland; and Quilt Patch, Fairfax, Virginia.

Second Printing

> Library of Congress Cataloging-in-Publication Data

Penders, Mary Coyne, 1931-
    Quilts from simple shapes : Triangles / Mary Coyne Penders.
       p.   cm.
    ISBN 0-913327-34-4 (pbk.) : $8.95
     1. Patchwork—Patterns. 2. Quilting—Patterns. 3. Triangle.
I. Title.
TT835.P449 1991
746.9' 7—dc20   91-35690
                                                          CIP

The Quilt Digest Press
P.O. Box 1331
Gualala, California 95445

ACKNOWLEDGMENTS

Books containing quilts to make include an array of measurements, templates and directions. Lenore Parham of Vienna, Virginia tackled these essential components with knowledge, skill and devotion to detail. In addition, whenever problems arose, Lenore was always available for consultation with a cheerful, immediate response. I am most grateful to Lenore not only for her expertise but also for her generous heart and her caring spirit.

Writing has always been easy for me, but ease evaporates as soon as I confront the intricacies of my new computer. Without the proficient services of my nineteen-year-old son Christopher, I would still be buried in the instruction manual. Many thanks to Chris for his ever-willing assistance and tutelage, as well as for the fun he contributed to our working together.

Special thanks are due Ardis and Robert James, prominent quilt collectors who generously shared their collection, making it possible to present antique quilts in this book.

My warmest thanks to Alex Anderson, whose creative talent is present in the two new quilts in this book. Alex received valuable assistance from Kristina Volker and Rosalie Sanders. For advice, information and encouragement, I am very grateful to the marvelous staff at The Quilt Patch in Fairfax, Virginia, and to Kay Lettau, who pieced the Special Project.

Acknowledgments are not complete without recognition of the editing and production skill of my wise editor, Harold Nadel, and the valuable advice and support of Jeff Bartee and Sharon Gilbert. To this outstanding team at The Quilt Digest Press, my heartfelt thanks for your services and your friendship.

Finally, to my husband Lee, who always understands, my love and thanks for your unfailing support and valuable counsel.

♦ ♦

For Harold and for Jeff, with abiding appreciation for their incomparable expertise and support.

INTRODUCTION

Triangles are a staple of the quiltmaker's diet. I'll bet you've consumed a lot of them, either visually by enjoying them in many treasured patterns and quilts, or by actually sewing them, perhaps in combination with other shapes, in your own quilts. In this book you'll find a feast of triangles: all the quilts are constructed primarily from this simple shape. *Quilts from Simple Shapes: Triangles* gives you a two-fold opportunity to make quilts that are easy as well as visually appealing.

When I began my quilt-teaching career working with beginners, I soon discovered that new students need easy techniques and freedom from complex designs in order to become familiar with the possibilities of color and fabric combinations. Then I made an interesting discovery about experienced quiltmakers who sought challenge in complicated patterns. Intricacy of design doesn't guarantee great results! Sometimes even seasoned quiltmakers need to take a break from complexity in order to focus on the crucial elements of color and fabric choices. Working with triangles gives you that freedom.

I can't count the number of times that a quilter has called me to lament that she is a year or two behind in finishing a quilt. "Mary, my daughter was born fifteen months ago and I still haven't finished her baby quilt. I'm afraid she'll be out of her crib and on her way to college before I get it quilted!" This is not outlandish when you consider that many quilters run a household, take care of a family, manage a career and volunteer in the community. Alas, all of these responsibilities cannot be set aside in favor of quiltmaking. For many people, quilting has to fit into the nooks and crannies of very busy schedules.

Have you ever needed to make a quilt in a hurry, perhaps for a wedding or a new baby or a graduation? I know that you don't want the finished product to appear hastily made or boring, and I also know that many of you, severely pressed for time but determined to quilt, are looking for solutions that do not sacrifice visual impact. *Quilts from Simple Shapes: Triangles* provides that solution with six quilts made from triangles. Three are heirloom quilts from the past, ready for your personal interpretation; two are sparkling new renditions of favorite patterns in today's fabrics. To help you with your gift list, I've included a small SPECIAL PROJECT that you can make in a jiffy. Study the color photographs and decide if you want to make one of the quilts exactly as shown. This is certainly tempting! I hope that as you become comfortable with triangles you may want to try your own color and fabric design ideas.

Whichever pattern you choose to make first, I know you'll appreciate the combination of making efficient use of your time and making a terrific quilt. Remember, all six quilts are simple designs that promise success with your triangle diet. Have a feast!

P.S. You'll love making the SPECIAL PROJECT in this book. Please be sure to include yourself on your gift list, and make or buy a doll who sleeps under your triangles.

Here's what *Quilts from Simple Shapes: Triangles* offers you:

◇ Hints to help you choose colors and fabrics
◇ Use of your scrap collection
◇ Templates in various sizes
◇ Yardage requirements
◇ Instructions for regular and quick cutting
◇ Directions for construction
◇ Teaching Plan

SUPPLIES

To make the quilts in this book, you will need:

Template plastic
Scissors or knife for cutting template plastic
8" Fabric scissors or rotary cutter and cutting mat
C-Thru plastic ruler, 2" x 18", or quick-cutting ruler
Fine-line pencil
Silver or white pencil for dark fabrics
Pencil sharpener
Glass-head pins
Seam ripper
Sewing machine or hand-sewing needle
Cotton sewing thread
Hand-quilting needle, between #8, #9 or #10
Quilting thread
Pressing surface
Iron
Batting
Quilting hoop or frame
Fabrics: yardage is given for each individual quilt
Containers (such as shoe boxes) for organizing fabric pieces
Good lighting and stable working surface

Pieced by Alex Anderson and machine-quilted by Patsi Hanseth.

BARN RAISING

Blue and white quilts are perennial favorites with quilters, so I wanted you to have an example in this book. I know from many years' teaching experience that most quilters accumulate more blue fabrics than any other color. I thought you'd like to use up some of your stash, or you might like to start from scratch as I did with a terrific new blue and white fabric.

This fabric is full of marine images: fish, shells, underwater foliage – it almost swims. What makes the fabric more interesting than most blue-and-whites is the fact that one of the blues is actually blue-violet, while another leans toward blue-green. While the overall effect is still blue and white, two colors closely related to blue lend subtle sparkle and pizazz. Several other fabrics, including blues from light to medium to dark, provide contrast against a white ground.

Alex and I had fun choosing fabrics to enhance the fish, including more shells, shallow water, pulsing waves, seamen's knotted rope and stars to steer by. The more we looked, the more we discovered fabrics that are compatible with a marine motif.

However, you don't need a motif fabric to make this quilt! Feel free to choose any colors or fabrics you wish, especially if you want to make a dent in your scrap bag with the large number of triangles in the *Barn Raising*. As long as you maintain the contrast between light and dark fabrics, your harmony will be successful.

HINT FROM MARY: *Contrast in value establishes pattern.* In quilts where one template shape is used, it is essential to distinguish between the shapes by means of alternating values. Value refers to the amount of light or dark in a color. When you are buying new fabrics, remember to include lights and dark with the more readily available medium values.

SIZE

This quilt measures 56" square. You can increase the dimensions by adding one or more borders. The size shown here is ideal for a large wall quilt or throw.

TEMPLATES

You will need one template: **T 1**.

YARDAGE

Over 35 different blue and white prints are used in this rendition of the *Barn Raising*. Of course you can choose any color harmony you wish, as long as you keep in mind that half of the triangles are light and the other half are dark. This quilt offers a good opportunity for dividing your scraps into lights and darks.

LIGHT PRINTS	2½ yards
DARK PRINTS	2½ yards
BACKING	3¼ yards
BINDING	⅔ yard

CUTTING

LIGHT PRINTS	T 1	784 triangles
DARK PRINTS	T 1	784 triangles

QUICK CUTTING

Divide your scraps into lights and darks. Stack the scraps and cut them into 2⅞" squares. Then cut the squares in half on the diagonal. (If you prefer to cut strips, use a width of 2⅞". Cut into 2⅞" squares, and then cut the squares in half on the diagonal.) You need a total of 784 light and 784 dark triangles.

CONSTRUCTION

1. The *Barn Raising* can be divided into 28 horizontal rows; each row has 28 squares. Each square is made from 2 triangles pieced together.

2. Follow the diagram for placement of light and dark fabrics. Sew pairs of triangles together to make 28 squares for Row 1.

3. Sew the squares together in a horizontal row.

4. Repeat for the remaining 27 rows, following the illustration carefully.

5. When you have completed the 28 rows, sew them together.

6. Finish with a simple binding as shown here, or add borders to increase the dimensions if you wish.

7. Prepare the finished quilt top for quilting. Press carefully, taking care not to distort the edges. Layer and baste the quilt top, batting and backing. Hand or machine quilt.

8. Finish the edges by attaching the binding.

QUILTING SUGGESTIONS

Here is the simplest of all quilting designs. Diagonal lines extend from the center to the four corners of the quilt, following the light-dark progression in the ditch or seam. This quilting design is highly suitable for machine quilting, which I recommend if you are making a quilt that will be subject to a lot of wear.

WILD GOOSE CHASE

Here's a wall quilt that suggests the migration of wild geese, flying across the sky against nature's background of fall foliage. Rich oranges, golds, rusts, greens and browns contribute to the climate of autumn. Alex has created lots of pattern and movement for a dramatic effect. If you prefer to make a tranquil quilt, you might try the budding colors of early spring, or the serene blues and greens of a summer day. This pattern is particularly well suited to scraps.

The large leafy fabric in the border is special. This fabric establishes the autumn mood of the quilt, introduces several colors which create the quilt's palette, and provides contrast between the larger design images and the small prints used for the geese. I like the visual surprise of the outline quilting around the leaves, which enhances both the individual fabric and the entire quilt.

HINT FROM MARY: *Look for variety in the scale of the fabrics you choose.* Scale refers to the size of the design images on the fabric. Some designs are very small, like microdots, while others range from small through medium to large, like the leafy print in the *Wild Goose Chase* border. Instead of placing fabrics of the same scale next to one another, it is more interesting to introduce variety in the size of printed images. If you look closely at the fabrics in this quilt, you can see how Alex used contrast of scale, in partnership with an intriguing array of visual images, to enrich the quilt.

SIZE

This quilt measures 49½" by 51½". Dimensions include a narrow ¾" dark red border and a wide 5" floral outer border. This size is perfect for a small wall quilt and is an ideal easy-to-make gift for a special occasion.

TEMPLATES

You will need two templates, **T 1** and **T 2**.

YARDAGE

Over 35 different prints were used for the geese, making this quilt a good scrap-quilt possibility. A large floral print is used between the rows of geese, and these sections are further defined by narrow red strips. Notice that the large-triangle geese in the quilt shown here are made from both light and dark fabrics, and the small background triangles are a mixture of lights and darks. Because the *Wild Goose Chase* pattern is also very effective when the large triangles are dark and the small triangles are light, the yardage and cutting directions given below follow this easy arrangement.

FLORAL PRINT	1½ yards for border strips and triangles
DARK RED STRIPS	1¼ yards for border strips and triangles
ASSORTED SCRAPS	1½ yards total for triangles, divided between lights and darks
BACKING	3¼ yards (2 lengths)
BINDING	⅔ yard

CUTTING

FLORAL PRINT	Cut 4 outside strips, 5½" wide, and 4 inside strips, 3½" wide, on the *lengthwise* grain of the fabric. Use some of the remaining fabric for geese triangles.
DARK RED STRIPS	Cut 12 border strips 1¼" wide on the *lengthwise* grain of the fabric. Use some of the remaining fabric for geese triangles.
ASSORTED SCRAPS	**T 2** cut 100 triangles–darks **T 1** cut 200 triangles–lights

Pieced by Alex Anderson and machine-quilted by Patsi Hanseth.

QUICK CUTTING

Follow cutting instructions above for the floral and the red strips. Add some leftover pieces to the geese scrap selection.

Stack your scraps and cut 2⅞" squares. You need 100 squares. Cut the squares in half on the diagonal to yield 200 triangles (light).

Stack your scraps and cut 5¼" squares. Cut the squares into quarters on the diagonal. You need 25 squares to make 100 triangles (dark).

CONSTRUCTION

The *Wild Goose Chase* quilt is made from units which are sewn into strips. A unit consists of one large (**T 2**) and two small (**T 1**) triangles. There are 20 units per strip. Five pieced strips of geese alternate with long strips of a large floral print. Narrow bands of dark red separate the geese units from the floral print strips.

1. Piece 100 individual units by sewing two **T 1** triangles to either side of one **T 2** triangle.

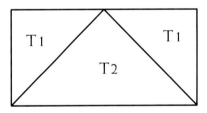

2. Sew 20 units into one strip. Make 5 of these strips.

3. Measure the length of the geese strips. Trim 8 of the narrow vertical red border strips, and 4 of the wide floral inside border strips, to the exact length of the geese strips.

4. Sew a narrow red strip to each side of a wide floral strip. Repeat 3 times until you have 4 complete sets consisting of floral strips bordered on either side by a dark red strip.

5. Lay the strips out, beginning with a geese unit on the left, and alternating geese with the red-plus-floral-plus-red units, ending with a geese unit on the right. Sew units together.

6. Measure the length and width of the assembled piece in order to determine the measurements for cutting the 4 remaining dark red borders.

7. Sew the narrow dark red border strips to each side of the quilt. Then attach the top and bottom strips.

8. Measure length and width again to determine measurements for trimming the wide outer floral borders to size.

9. Sew the sides first, and then attach the top and bottom.

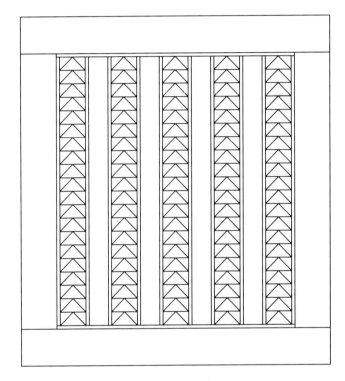

10. Prepare the finished quilt top for quilting. Press carefully, taking care not to distort the edges. Layer and baste the quilt top, batting and backing. Hand or machine quilt.

11. Finish the edges by attaching the binding.

QUILTING SUGGESTIONS

The geese are outline-quilted around each triangle, with the stitches in the ditch or seam. The large autumnal print used for the strips and borders is beautifully quilted with stitching around each leaf, emphasizing the contours of the foliage. Meandering quilting in the brown background areas of the leaf print further enhances the border quilting.

Amish maker unknown, c. 1930-1950. Collection of Robert and Ardis James.

OCEAN WAVES

What a fabulous quilt! Black is usually a guarantee of success with a color harmony, and in this lovely quilt black does not disappoint us. Combined with pastels, black is a dramatic contrast. How vivid the colors of the waves appear as they float above the inky depths of the ocean.

This rendition of the popular *Ocean Waves* is made from solid fabrics in an array of eleven colors, including values of purple, green, pink, yellow and blue. Notice how the triangles are arranged in a light-dark alternating pattern. Sometimes the value pattern is changed, which leads the eye to random places where two similar values are side-by-side.

I especially like the containment of this quilt. The borders enclose the field of the quilt in a positive way. The inner border is an enclosure of brightness, unifying the light elements of the patchwork within a frame. The wide, dark outer border surrounds the design with dramatic emphasis.

HINT FROM MARY: *Borders are a very important part of your quilt.* It is really crucial to the success of a quilt to develop a border or borders that serve as a definitive frame. Whether plain or fancy, the border should be an integral part of the whole. Sometimes a border seems to be an add-on, unrelated to the quilt top; or the border may appear to be dropping off into outer space. When you finish a quilt top, you may want to try several colors or fabrics to see which one frames your quilt and enhances the overall appearance of the design. Study quilts in books and magazines to find inspiration for borders. Be sure to take the time to plan for this very important part of your quilt.

SIZE

This quilt measures 60" by 80". Dimensions include a 2" light blue inner border and an 8" black outer border.

TEMPLATES

You will need two templates, **T 1**, **T 3**.

YARDAGE

The antique *Ocean Waves* shown here is made from fourteen fabrics. You might choose a different color harmony; you could also look into your scrap bag for small-triangle fabric possibilities.

ASSORTED COLORS	¼ to ⅓ yard each of these solids: lavender, purple, light green, medium green, dark green, light pink, medium pink, gray, light yellow-gold, medium yellow-gold and dark blue (total: 11 colors)
LIGHT BLUE	1¾ yards (includes inner border)
MEDIUM BLUE	⅔ yard (includes binding and triangles)
BLACK	2⅝ yards (includes border)

CUTTING

ASSORTED COLORS	**T 1**	720 triangles
LIGHT BLUE		Cut 4 border strips 2½" wide on the *lengthwise* grain of the fabric.
MEDIUM BLUE	**T 1**	48 triangles
BLACK		The following order of cuts is important for fitting all the pieces into the yardage.
	T 3	30 triangles. From remaining fabric, cut 4 border strips 8½" wide on the *lengthwise* grain of the fabric. Then use **T 1** for 18 additional triangles.

QUICK CUTTING

ASSORTED COLORS — Cut strips 2⅞" wide. Cut them into 2⅞" squares. Cut squares in half on the diagonal. Or, if you prefer, stack scraps and cut 2⅞" squares, then cut them in half on the diagonal. You need a total of 720 triangles.

LIGHT BLUE — Cut 4 border strips 2½" wide on the *lengthwise* grain of the fabric.

MEDIUM BLUE — Cut three strips 2⅞" wide. Cut into 2⅞" squares. Cut squares in half on the diagonal for a total of 48. Reserve remaining fabric for binding.

BLACK — The following order of cutting is important for fitting all the pieces into the yardage.

Cut three 6⅞"-wide strips on the *crosswise* grain of the fabric. Cut into 6⅞" squares.

Cut the squares in half on the diagonal for a total of 36 triangles.

Cut four border strips 8½" wide on the *lengthwise* grain of the fabric.

Cut six 6⅞" strips. Cut into 6⅞" squares. Cut the squares in half on the diagonal for a total of 12 triangles.

CONSTRUCTION

A total of 24 ten-inch blocks are set together 4 x 6. There is a 2" light blue inner border and an 8" black outer border.

1. There are 32 small triangles in each block, arranged in 4 diagonal rows. Lay out your triangles according to the diagram, experimenting with placement until you arrive at a pleasing arrangement.

2. Sew the small triangles in diagonal rows, and then sew the row together.

3. Attach the 2 larger triangles to each side to complete the block.

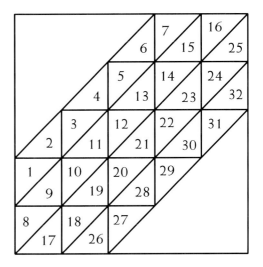

4. Sew the blocks together.

5. Measure the length and width of the completed quilt top to determine the measurements for trimming border strips to fit.

6. Trim two light blue inner border strips to the required length measurement and attach to the sides of the quilt top. Then trim and add the top and bottom strips.

7. Repeat the process, first measuring the length and width, and then trimming the black outer border to fit.

8. Sew the sides first and then attach top and bottom.

9. Prepare the finished quilt top for quilting. Press carefully, taking care not to distort the edges. Layer and baste the quilt top, batting and backing. Hand or machine quilt.

10. Finish the edges by attaching the binding.

QUILTING SUGGESTIONS

The small triangles are quilted in close proximity to the seam line. The stitching, which follows the long diagonal of the pattern, is no more than one-eighth inch from the seam. The large black squares feature echo quilting beginning with a small square in the center, and continuing with squares that fill the block out to the edge. The narrow blue border features a single-strand cable, while the wide black border is replete with a classic cable of several strands.

Maker unknown, c. 1875-1900. Collection of Robert and Ardis James.

THOUSAND PYRAMIDS

This pattern is among the best-loved of triangle designs, perhaps because it evokes images of the scraps of everyday life sewn into the comfort of a quilt. While there are many quilters who have fabric pieces from their wedding and maternity and children's clothing, enabling them to make truly meaningful *Thousand Pyramids*, I suspect there are even more of us who have baskets and boxes of scraps from other sewing or quilting projects, just waiting to be liberated by this pattern.

When you study the quilt, you will notice that contrast is maintained by the use of light and dark fabrics. Some of the darks are quite dark, while others are more medium-dark to medium. All of the fabrics in the medium to dark category are placed next to very light or light fabrics, creating the contrast that establishes the pattern. Value, or the amount of light or dark in a color, is very important when you want design elements to be distinct from one another.

The beauty of this quilt, aside from the loveliness of the colors themselves, is in the arrangement of values. Rather than a rigid arrangement of light and dark, we find a surprising variety of subtle contrasts. While some darks have a very pale light next to them, which causes the dark to appear even darker, other darks are placed next to a medium-light, so that they are not quite so dark in that portion of the quilt.

Bright blue is scattered sparingly through the quilt, providing an unexpected contrast with the dull hues of brown. But the main contrast comes from what my friend Lenore immediately termed "bubblegum pink" when she saw this quilt. This rich, chewy kind of pink is such a wonderful foil for the somber tans, browns and grays. Think about light-dark and bright-dull contrasts when you design your quilt.

HINT FROM MARY: *The value of a color or fabric depends on what is placed next to it.* After you have cut out some of the *Thousand Pyramids* triangles, try them out in various value positions. Take a medium and make it appear darker by placing the lightest light next to it; make it appear lighter by placing the darkest dark next to it. Keep experimenting until you discover more combinations that illustrate how value can be changed.

You can use this knowledge to create a spontaneous feeling in your quilts, as opposed to the rigid feeling that proceeds from always using fabrics in a designated value category. Manipulate your fabrics! You'll enlarge the scope of your collection and your quilts will become livelier.

SIZE

Once in a while, the width of an antique quilt is greater than its length, which you see in this example measuring 76" wide and 72" long. I've included templates for sizes more suitable for contemporary beds. Each *Thousands Pyramids* quilt, regardless of size, is constructed exactly the same way. Select the overall measurement that best suits your needs, and then use the templates indicated for that size. You will need two templates, one for the pyramids and one for finishing the sides.

QUILT SIZE	TEMPLATE	TRIANGLE SIZE
64" x 72"	**T 9, T 10**	Base 2" Height 2¼"
72" x 72"	**T 11, T 12**	Base 2¼" Height 2¼"
80" x 80"	**T 13, T 14**	Base 2½" Height 2½"
80" x 88"	**T 15, T 16**	Base 2½" Height 2¾"

TEMPLATES

Select your templates for the size you have chosen to make (as indicated above) from the following list. You will need two triangles: one for the pyramids, and one for finishing the sides.

1) **T 9**, **T 10**
2) **T 11**, **T 12**
3) **T 13**, **T 14**
4) **T 1**, **T 16**

YARDAGE

To make any of these quilt sizes, you will need a large assortment of prints totaling from 8 to 10 yards. Each ¼ yard of fabric yields from 50 to 80 triangles, depending on the size of your template. The sizes given here require approximately two thousand triangles or more, which should deplete your scrap bag! The amounts given below will make any of the quilt sizes described above. If you change the dimensions, simply cut more triangles as needed.

The quilt in the illustration displays a wide variety of colors and visual images. Black, brown, tan and beige are featured, with a few blues and a rich "bubblegum" pink providing contrast. After choosing your color scheme, search for many different fabrics in each color so that you can add visual interest to your composition. Divide your fabrics equally between lights and darks.

BACKING	2 lengths for quilts up to 80" wide 3 lengths for quilts over 80" wide
BINDING	⅔ yard for quilts up to 80" wide ⅞ yard for quilts over 80" wide

CUTTING

The following instructions apply to each set of two templates. Stack your fabrics and cut several layers at one time. Once the cutting is done, you're home free and the rest is easy! After you have cut the 64 side triangles used to fill in the sides of the quilt, put them aside until needed, so you don't confuse them with the pyramid triangles.

PYRAMID TRIANGLES	Cut 2016 (half light, half dark).
SIDE TRIANGLES	Cut 16 light and 16 dark from the side template. Then *reverse* the template and cut 16 light and 16 dark. You need a total of 64.

CONSTRUCTION

1. Piece all units as shown. One unit consists of 16 triangles. Piecing in units rather than rows makes it easier for you to make corrections and adjustments to the overall design as you go along.

2. Unit #1 has a dark triangle at the top, while Unit #2 has a light triangle at the top. Piece 120 units: 60 of Unit #1 and 60 of Unit #2, according to the diagram.

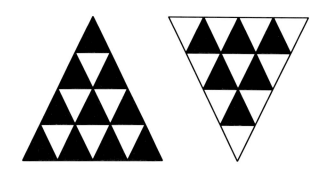

3. Focus on each individual unit, one at a time, as you make it. Consider the contrasts of color, of visual design on the fabrics, and of light and dark values.

4. Piece all the units before assembling them into a quilt. From time to time, lay the completed units out. Place Unit #1 (with the dark triangle at the top) facing upward. Place Unit #2 (with the light triangle at the top) facing downward. Try different arrangements; you can add or distribute colors as needed. Experiment until you are pleased with the results.

5. Unit #3, Unit #4, Unit #5 and Unit #6 are used to fill in the sides. Piece 8 of each unit as illustrated. Refer to the illustration for placement of the side units.

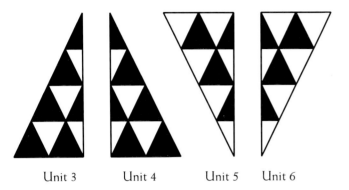

Unit 3   Unit 4   Unit 5   Unit 6

6. Assemble all the completed units together as illustrated.

7. Prepare the finished quilt top for quilting. Press carefully, taking care not to distort the edges. Layer and baste the quilt top, batting and backing. Hand or machine quilt.

8. Finish the edges by attaching the binding.

QUILTING SUGGESTIONS

Because of the large number of small pieces and the busyness of the pattern, *Thousand Pyramids* is a good candidate for very simple quilting. Quilting stitches follow the diagonal lines of the pattern to produce cross-hatch stitching that is functional rather than decorative.

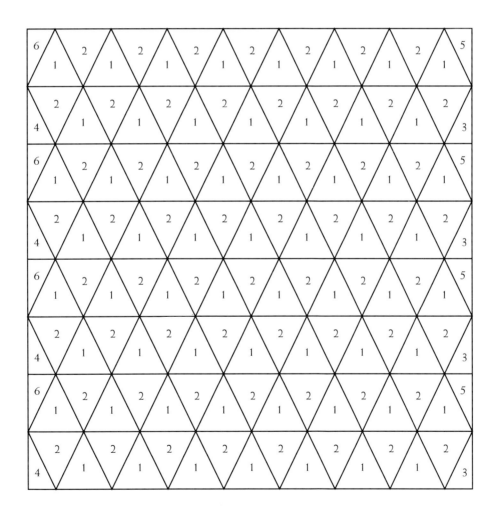

Maker unknown, 1920. Collection of Robert and Ardis James.

VIRGINIA REEL

Are you ready to put on your dancing shoes? Just look at the spinning motion in this quilt! I'm partial to this pattern because I live in Virginia, but I'm also attracted by the crisp red, white and blue color scheme. It's so very American, a harmony for quilters who want to make a patriotic quilt.

The *Virginia Reel* is also the perfect pattern for those who love beautiful quilting. The diagonal set provides plain blocks which need to be embellished with fine quilting stitches. I think this quilt is an appealing gift to thrill the hearts of any bride and groom or special anniversary couple. Or, if you've been making gifts for others all your life, but never for yourself, why not treat yourself and your home to this lovely tradition from America's past?

HINT FROM MARY: *Elegant quilting enhances traditional quilts.* Don't stint. Find a design that fills the space. Use matching thread. If you've never quilted, find a class or a book to help you, and practice on a test square instead of your quilt until you are comfortable with the quilting stitch. Uniformity or consistency is more important than size, so don't worry about getting an astronomical number of stitches to the inch. Strive instead for evenness. Relax and enjoy the process!

SIZE

The finished size of this quilt is 70½" square. This dimension includes three 1¾" borders of blue, muslin and red. There are 25 pieced blocks, set 5 x 5 on the diagonal, and 16 alternate blocks in the antique quilt shown here.

TEMPLATES

You will need four templates: **T 3**, **T 5**, **T 6**, **T 7**.

YARDAGE

The antique quilt shown here uses three solid fabrics.

BLUE	2 yards (includes inner border)
RED	2¼ yards (includes outer border)
MUSLIN	4¼ yards (includes middle border)
BACKING	4½ yards (2 lengths)
BINDING	⅝ yard

CUTTING

You are going to cut three sets of borders; they will be trimmed to size after the blocks are sewn together and an accurate measurement is made.

BLUE — Cut 4 border strips 2¼" wide on the *lengthwise* grain of the fabric. Then cut template pieces from the remaining fabric.

T 6   100 triangles (large)

RED — Cut 4 border strips 2¼" wide on the *lengthwise* grain of the fabric. Use the remainder of the fabric to cut template pieces.

T 5   200 triangles (small)

MUSLIN — Cut 4 border strips 2¼" wide by 72" long on the *lengthwise* grain of the fabric.

Use the rest of the fabric for the following templates:

Cut 16 squares 9½" x 9½" for the alternate blocks.

T 5   200 triangles (small), also cut across the full width (*crosswise*) of the fabric.

T 6   100 triangles (large).

T 7   16 side triangles. Place the *long* side of the triangle on the straight of the grain.

T 3   4 corner triangles. Place the *short* side of the triangle on the straight of the grain.

QUICK CUTTING

BLUE
Cut 4 border strips 2¼" wide on the *lengthwise* grain of the fabric. From the remainder, cut 5 rows of 5½"-wide strips on the *crosswise* grain. Cut the strips into 5½" squares. Cut the squares into 4 quarters on the diagonal, making 100 triangles.

RED
Cut 4 border strips 2¼" wide on the *lengthwise* grain of the fabric. From the remainder, cut 10 rows of 3"-wide strips on the *crosswise* grain. Cut the strips into 3" squares. Cut the squares in half on the diagonal, making 200 triangles.

MUSLIN
Cut 4 border strips 2¼" wide by 72" long on the *lengthwise* grain of the fabric. From the remainder of the width left from cutting the borders, cut 10 rows of strips 3" wide on the *crosswise* grain. Cut the strips into 3" squares, making a total of 200.

Cut 6 rows of 5½" strips. Cut the strips into 5½" squares. Cut the squares into quarters on the diagonal, making 100 triangles.

Cutting across the entire width of fabric, cut 4 rows of strips each 9" wide. Cut the strips into 9 squares, making a total of 16 alternate blocks.

Cut 2 rows of 13¼"-wide strips. Cut the strips into 13½" squares. Cut the squares into quarters on the diagonal, making 16 side triangles.

Cut two 6⅞" squares. Cut the squares in half on the diagonal, making 4 corner triangles.

CONSTRUCTION

1. Piece 100 sets of Unit #1, sewing two muslin triangles (**T 6**) to one blue triangle (**T 5**).

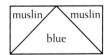

2. Piece 100 sets of Unit #2, sewing two red triangles (**T 6**) to one muslin triangle (**T 5**).

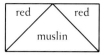

3. Sew one Unit #1 and one Unit #2 together to form Unit #3.

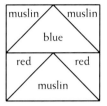

4. Sew four of Unit #3 together to make one complete block. Piece twenty-five blocks.

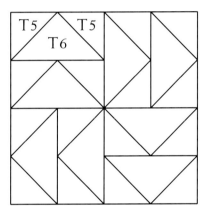

5. Assemble the pieced blocks with the alternate blocks in rows as illustrated, beginning with corner or side triangles in order to make a complete row.

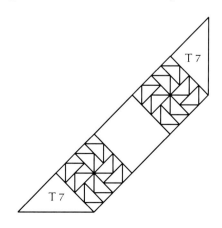

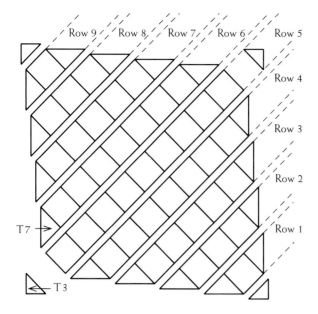

6. Measure the length of the quilt top. Trim two blue border strips to fit these measurements. Sew to each side.

7. Measure the width of the quilt top. Trim the two remaining blue border strips to fit these measurements. Sew to the top and bottom.

8. Measure, trim and sew the muslin border strips in the same order: sides first, followed by top and bottom.

9. Add the final border, the red strips, in the same order. Always complete one border before adding another.

10. Prepare the finished quilt top for quilting. Press carefully, taking care not to distort the edges. Layer and baste the quilt top, batting and backing. Hand or machine quilt.

11. Finish the edges by attaching the binding.

QUILTING SUGGESTIONS

Here is a treat for lovers of beautiful quilting. Every area of this *Virginia Reel* is embellished with fine stitching, beginning with the tulip design in the muslin blocks. All of the areas in and around the flowers, buds and stems are filled with stipple quilting. In the pinwheel portions of the pattern, the spaces between the pinwheel blades have been stippled with a geometric design. Indeed, all of the background areas throughout the entire quilt are heavily quilted. Around the borders, diagonal lines of quilting no more than one-quarter inch apart extend in one direction out to the edge.

ADDITIONAL SIZES

The *Virginia Reel* may also be made be made in the following sizes, all with the same 8½" block.

70½" x 82½", with a 10½" border: make 30 pieced blocks, set 5 x 6, and 20 alternate blocks. Use 18 side and 4 corner triangles.

82½" x 94½", with a 10½" border: make 42 pieced blocks, set 6 x 7, and 30 alternate blocks. Use 22 side and 4 corner triangles.

94½" square, with a 10½" border: make 49 pieced blocks, set 7 x 7, and 36 alternate blocks. Use 24 side and 4 corner triangles.

YARDAGE FOR ADDITIONAL SIZES

	70½" x 82½"	82½" x 94½"	94½" square
BLUE	2¼ yards	2½ yards	2¾ yards
RED	2⅜ yards	2⅔ yards	2⅞ yards
MUSLIN	4¾ yards	5 yards	5¼ yards
BACKING	5 yards	5½ yards	8¼ yards
BINDING	¾ yard	¾ yard	⅞ yard

Pieced by Kay Lettau and machine-quilted by Alex Anderson.

Special Project
DOLL'S KALEIDOSCOPE

This small project is sure to delight all the little girls in your life. Because my daughter never played with dolls, I've been absent from this milieu for a very long time. I had so much fun designing this quilt! Childhood memories of Sophranesba, my favorite doll, kept creeping into my consciousness; I just knew she would love pink and green and yellow. So this little quilt is for Sophranesba, who resides now only in memory.

This is a sweet quilt, with images of pale butterflies and bright calico flowers. There are two yellow prints and several greens, one a nice plaid contrast. I especially like the muted print that looks almost like a solid, used for the corner triangles in each block. But Sophranesba was not all sweetness (she had quite a temper), so I dug out a wonderful pink and white check to put some punch in the border.

Alex quilted this piece, outlining the kaleidoscope triangles and placing hearts and leaves in the border. We think it's darling and hope you'll make one for everyone you know who ever played with dolls!

HINT FROM MARY: *Small is beautiful!* You probably have a gift list that's way too long, and you'd like to make handmade presents for family and friends. Why not finally accomplish this goal by going small? This little doll quilt can be made easily and speedily. When you finish covering the family dolls (don't forget to include nostalgic types like me, who will love our dolls right into old age), you can use the same pattern to make many other gifts as well.

Four kaleidoscope blocks form a place mat with the addition of borders. Double the size of the quilt shown here; then add a second border to make a card-table cover. Enlarge the doll quilt by adding enough blocks to cover a baby's crib (or enlarge the size of the block). Double the quilt horizontally to make a table runner.

SIZE

This small quilt measures 15½" x 21½", including the 1¾" border. There are six 6" blocks.

TEMPLATES

You will need 2 templates, **T 4**, **T 8**.

YARDAGE

Use scraps! I went into my stash and found a light pink butterfly and a medium pink calico flower. There are two yellows, similar in value, but one is brighter than the other for contrast. I decided on three greens: one is a soft floral with pink and yellow flowers; the second is a slightly darker leaf and berry print; the third is a plaid, for visual texture contrast. A soft yellow and pink floral appears only in the border corners, while a marvelous colorwash print functions as a solid in the corners of each *Kaleidoscope* block.

ASSORTED PRINTS	½ yard total for triangles
BORDER	¼ yard
BACKING	½ yard (includes the binding)

CUTTING

ASSORTED CORNER TRIANGLES	**T 4**	32
ASSORTED KALEIDOSCOPE TRIANGLES	**T 8**	48
BORDER		Cut 2 strips 2¼" wide on the *crosswise* grain of the fabric.

QUICK CUTTING

Stack your fabrics. Then, using template **T 4**, cut several layers at once with your rotary cutter until you have a total of 32 triangles. Use template **T 8** to stack-cut 48 triangles. Follow directions above for cutting the border.

CONSTRUCTION

If you are making the quilt as shown, you need to make six blocks. Each block consists of 8 large and 4 small triangles.

1. Sew large triangles in pairs, as shown in the diagram. Then sew 2 pair together to make half of the block.

2. Sew the small corner triangles to each half.
3. Sew the two halves together to complete the block, being careful to match the centers. Repeat the process until all the blocks are made.

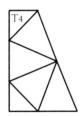 

4. Sew completed blocks together.
5. Measure the length and width to determine exact measurements for trimming border strips.
6. Cut side borders to fit and attach to the quilt top.
7. Sew the border corner triangles together to make corner squares.
8. Again, measure the width of the quilt, and trim the border strips to fit. Sew corner squares to each end of the top and bottom border strip

9. Finally, attach the top and bottom borders.
10. Prepare the finished quilt top for quilting. Press carefully, taking care not to distort the edges. Layer and baste the quilt top, batting and backing. Hand or machine quilt.
11. Finish the edges by attaching the binding.

QUILTING SUGGESTIONS

Traditional quilting contributes to the charm of this small quilt. Each of the kaleidoscope triangles is quilted on two long sides, one-quarter inch from the seam allowance. Where four corner triangles come together to form a square, the quilting design echoes the square shape, with the stitches again one-quarter inch from the seam line. The borders display hearts in the center of each strip, with simple feathers continuing out to the ends. This is a nice project to carry with you for hand-quilting; it's also very durable when quilted by machine, a consideration if the quilt is earmarked for an active playroom.

TEACHING PLAN

The three books in this series present an excellent opportunity for quilting teachers and shop owners. Each individual book is suitable for a series of six lessons. Offering courses from all three books, *Squares*, *Triangles* and *Rectangles*, provides your students with consecutive lessons spanning three seasons of the year. Each student could expect to produce three quilt tops in this period of time (for example, from fall, winter and spring courses). You may wish to include a copy of the book in the tuition price for the course as a means of attracting students.

Preceding the course outline, you will find a structure for individual lessons. Using a lesson plan enables you, the teacher, to organize class time well; it also fosters learning because students respond enthusiastically to classes that are well presented. I hope you'll incorporate your own creative ideas into this plan.

SUGGESTED STRUCTURE FOR THE LESSONS

1. Introduction of lesson content
2. Presentation of supplies
3. Discussion of color and fabric theory
4. Demonstration of techniques
5. Group and individualized instruction
6. Evaluation and constructive criticism
7. Assignment of homework

After the course is underway, I recommend incorporating Show and Tell into every lesson. This is an enjoyable way to give recognition, inspiration and encouragement to students.

FORMAT FOR A SIX-LESSON COURSE

LESSON ONE — Emphasis on selection of quilt by each student OR assignment of quilt by the teacher if you prefer to teach one quilt to the entire class. Numbers 1, 2, 3 and 4 from the lesson structure are important parts of the first class.

LESSON TWO — Emphasis on in-store fabric-buying lesson, with selection of fabrics for each quilt. Planning color/fabric placement. Instruction for fabric preparation and cutting techniques. This is a busy class, but you'll still have time for numbers 1, 3, 5 and 7.

LESSON THREE — Emphasis on demonstration of techniques for machine or hand piecing. Include numbers 1 through 7.

LESSON FOUR — Emphasis on problem-solving and constructive criticism. Again, numbers 1 through 7 provide a good checklist.

LESSON FIVE — Emphasis on techniques for the addition of borders. Use the lesson structure as you introduce the new element of borders.

LESSON SIX — Emphasis on marking the top for quilting and making the binding. Summarize the main points of the previous lessons, and pay particular attention to numbers 5 and 6.

Schedule a Class Reunion so that you and your students may enjoy the finished quilts together!

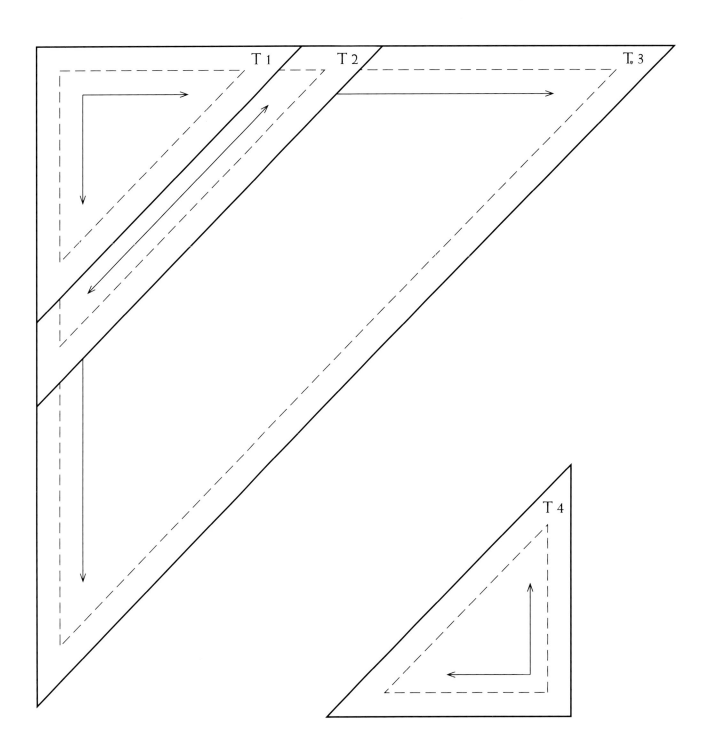

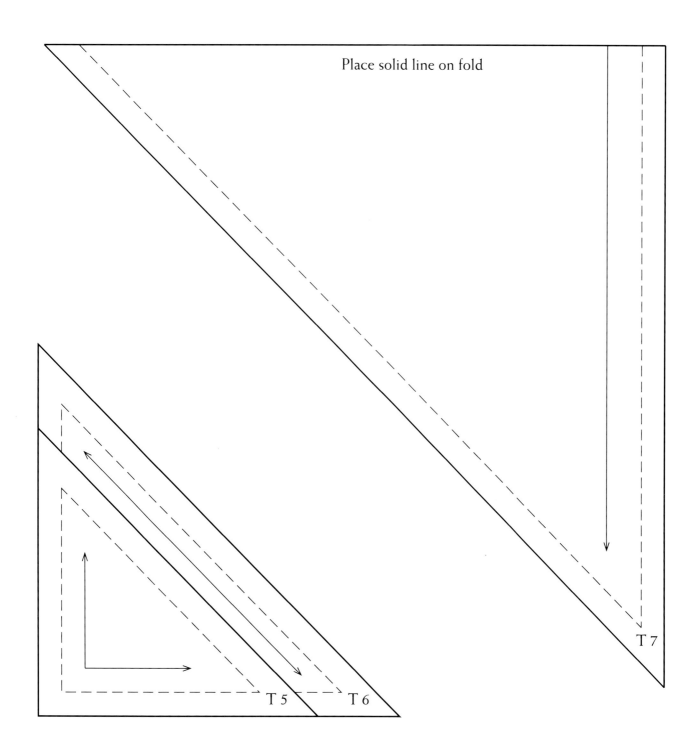

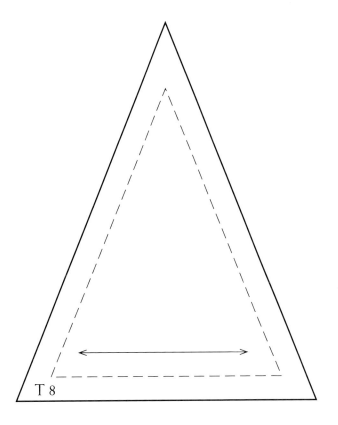
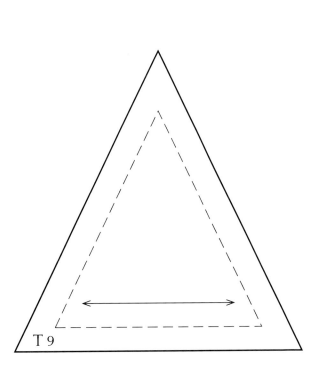
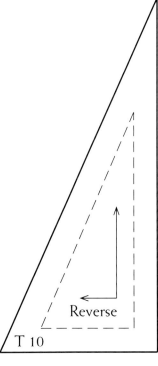

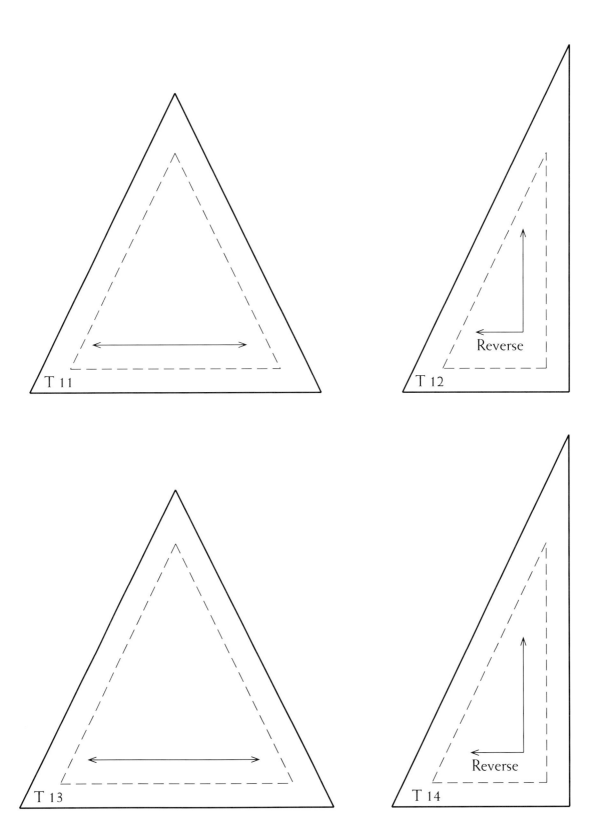

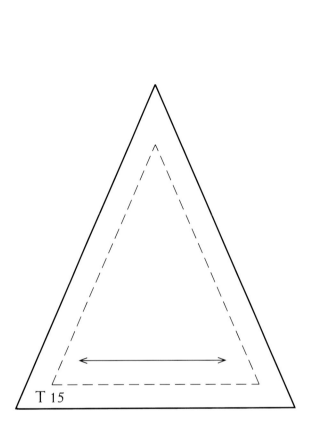
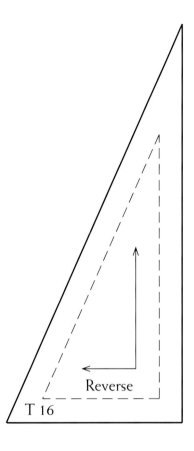